"Footprints of Change:

A Guide to Minimizing Our Impact on the Environment"

By

Iram Nazir

Short Summary

"Footprints of Change: A Guide to Minimizing Our Impact on the Environment" is a book that explores the impact of human actions on the environment. The book discusses the ways in which our daily habits and choices contribute to environmental degradation and highlights the need for individuals and society as a whole to take responsibility for the consequences of our actions. The book provides a comprehensive overview of the most pressing environmental issues of our time and offers practical solutions for reducing our footprint and preserving the planet for future generations.

Chapter 1: Introduction

The impact of human actions on the environment is a topic of increasing concern and importance in today's world. With the rapid pace of industrialization and urbanization, the footprint of human activity is becoming increasingly apparent, both in terms of the damage it is causing to the planet, and in terms of the consequences of these actions for future generations.

As a result, it has become increasingly important to understand the nature of this impact, and to explore ways in which it can be mitigated and reduced. This book is dedicated to exploring the topic of the human footprint, and provides a comprehensive and accessible examination of the environmental consequences of our actions.

In this chapter, we will provide an overview of the key points that will be discussed throughout the book, including the causes and effects of human activity on the environment, the ways in which these impacts are being measured and monitored, and the steps that can be taken to reduce the human footprint and promote a more sustainable future. Whether you are a concerned citizen, a policy maker, or a student of environmental science, this book will provide valuable insights and guidance as you navigate the complex and challenging landscape of the human footprint.

Chapter 2: The Science Behind Environmental Consequences

The environmental impact of human activities has become a critical issue in recent years, with the science behind these consequences becoming increasingly well understood. In this chapter, we will examine some of the key areas in which human activities have had an impact on the environment, including climate change, deforestation, air and water pollution, and loss of biodiversity.

Climate Change: One of the most significant environmental consequences of human activities is climate change. Climate change refers to the long-term change in the average weather patterns that have come to define Earth's local and global climates. The science behind climate change is rooted in the understanding that the Earth's climate is primarily driven by the balance of incoming and outgoing energy from the sun. This balance is controlled by several factors, including the Earth's atmosphere and its composition, the presence of greenhouse gases, and changes in the Earth's albedo.

The main cause of climate change is the emission of greenhouse gases, such as carbon dioxide, into the atmosphere. These gases trap heat from the sun, causing the Earth's temperature to rise. This rise in temperature is causing the melting of polar ice caps and glaciers, which contributes to rising sea levels, and the intensification of weather patterns, such as hurricanes, droughts, and floods.

Deforestation: Another significant environmental consequence of human activities is deforestation. Deforestation is the clearing of forests for agriculture, urbanization, or other uses. The loss of forests has a significant impact on the environment, including changes in the Earth's climate, soil erosion, and loss of biodiversity.

Forests play a crucial role in the Earth's climate by absorbing carbon dioxide from the atmosphere and storing it in the form of biomass. When forests are cleared, this carbon is released back into the atmosphere, contributing to the buildup of greenhouse gases and the acceleration of climate change. In addition, deforestation leads to soil erosion, as the roots of trees help to hold the soil in place. Finally, the loss of forests leads to the loss of biodiversity, as many species of plants and animals are only found in specific forest ecosystems.

Air and Water Pollution: Human activities also have a significant impact on air and water quality. Air pollution is caused by the emission of harmful substances into the atmosphere, such as sulfur dioxide, nitrogen oxides, and particulate matter. These substances contribute to a range of negative health effects, including respiratory and cardiovascular disease, and can also damage crops and forests.

Water pollution is caused by the release of pollutants into waterways, including chemicals, sewage, and other waste products. This pollution can harm aquatic life, disrupt ecosystems, and pose a threat to human health. In addition, water pollution can make it difficult or impossible to use waterways for activities such as fishing, recreation, and drinking water supply.

Loss of Biodiversity: Finally, human activities have led to the loss of biodiversity, or the variety of plant and animal species in the world. Biodiversity is important for maintaining the balance of ecosystems, providing ecosystem services, and supporting human well-being. However, human activities such as deforestation, climate change, and pollution have led to the decline of many species, and the loss of biodiversity can have negative impacts on ecosystems and human societies.

In conclusion, the environmental consequences of human activities are far-reaching and have significant impacts on the health of the planet and its inhabitants. The science behind these consequences is well understood and highlights the need for immediate action to mitigate the negative impacts of human activities on the environment. By understanding the science behind these consequences, we can work towards finding solutions and reducing the impact of human activities on the

Chapter 3: The History of Human Impact on the Environment

Human activities have had a profound impact on the environment for centuries, and a closer examination of the history of human impact on the environment can help us to understand the current state of the planet and the challenges we face today.

The earliest civilizations, such as the Ancient Egyptians, Greeks, and Romans, had a significant impact on their local environments. For example, the Ancient Egyptians built massive structures like the pyramids, which required the extraction of large amounts of stone and minerals from the surrounding landscape. This caused damage to the ecosystem, as well as soil erosion and the depletion of natural resources.

During the Middle Ages, deforestation was a common problem in Europe, as people cut down trees for fuel and building materials. This had a major impact on the environment, as forests play a crucial role in regulating the water cycle and supporting biodiversity.

As human civilizations expanded, so did their impact on the environment. The Industrial Revolution, which began in Britain in the late 18th century, saw a massive increase in industrial production, as well as significant advances in transportation and communication. This led to significant environmental consequences, including air and water pollution, the destruction of natural habitats, and the decline of biodiversity.

One of the most significant events in the history of human impact on the environment was the widespread use of synthetic chemicals, which began in the mid-20th century. Chemicals such as pesticides, herbicides, and fertilizers were developed to help improve food production and control pests and diseases. However, these chemicals have had a profound impact on the environment, contaminating soil and water, killing off beneficial insects and birds, and altering the balance of ecosystems.

The 20th century also saw the rise of the modern environmental movement, as people became increasingly aware of the impact of human activities on the environment. In 1962, Rachel Carson's groundbreaking book "Silent Spring" exposed the dangers of synthetic chemicals and helped to galvanize the environmental movement. This led to the creation of environmental laws and regulations, as well as the establishment of environmental organizations and advocacy groups.

Today, human impact on the environment continues to be a major concern. Climate change, caused by the emission of greenhouse gases from human activities, is one of the biggest challenges facing the planet. Climate change is causing sea levels to rise, leading to flooding and erosion, as well as altering weather patterns and causing severe weather events.

In conclusion, the history of human impact on the environment is a long and complex one, marked by both progress and destruction. Understanding this history is crucial if we are to address the environmental challenges facing the planet today and ensure a sustainable future for generations to come.

Chapter 4: The Human Footprint in Different Regions and Industries

Part 1

Human activities have had a profound impact on the environment over the centuries. The effects of this impact can be seen in different regions of the world, as well as in various industries. This chapter will examine the different types of human activities that have the greatest impact on the environment, including agriculture, transportation, energy production, and urban development. We will also look at the environmental impact of human activities in different regions of the world, including both developed and developing countries.

Agriculture is one of the oldest and most important industries in the world. It is responsible for producing the food that feeds the global population, but it also has a significant impact on the environment. Agricultural activities, such as plowing, tilling, and irrigation, can disrupt the delicate balance of ecosystems, leading to soil degradation, loss of biodiversity, and increased erosion. In addition, the use of pesticides and fertilizers can have a negative impact on air and water quality, leading to the contamination of local ecosystems and the death of wildlife.

Transportation is another key industry that has a significant impact on the environment. The burning of fossil fuels to power vehicles produces greenhouse gases, which contribute to climate change. In addition, the construction of roads, bridges, and highways can disrupt natural habitats, leading to the displacement of wildlife and the loss of biodiversity. The transportation sector is also responsible for producing significant amounts of air and water pollution, which can have negative effects on the health of both people and the environment.

Energy production is another major contributor to environmental degradation. The burning of fossil fuels to generate electricity produces significant amounts of greenhouse gases, including carbon dioxide, which contributes to climate change. In addition, the production and use of nuclear energy can lead to the release of radioactive materials into the environment, which can have long-term consequences for the health of people and the environment. The use of renewable energy sources, such as wind and solar power, is becoming increasingly popular as a way to reduce the environmental impact of energy production, but it still has a long way to go to replace traditional energy sources.

Urban development is another major contributor to environmental degradation. The construction of buildings, roads, and other infrastructure can disrupt natural habitats and lead to the displacement of wildlife. In addition, the growth of cities can contribute to air and water pollution, as well as the generation of large amounts of waste. The urbanization of many regions of the world has also led to the loss of green space and the increase of heat islands, which can have negative impacts on local ecosystems and the health of people.

The impact of human activities on the environment can be seen in different regions of the world. In developed countries, the environmental impact is often driven by the pursuit of economic growth and the use of industrial technologies. In developing countries, the impact is often driven by poverty, lack of access to clean water and sanitation, and the need to meet basic needs such as food, shelter, and energy. In many regions of the world, the impact of human activities on the environment is a complex interplay between economic, social, and political factors.

The environmental impact of human activities in different regions of the world is not uniform. In developed countries, the impact is often greater in terms of the scale of the damage and the long-term consequences. In developing countries, the impact is often greater in terms of the severity of the effects on people and the environment. For example, the deforestation of the Amazon rainforest has a greater impact on the environment in terms of the loss of biodiversity and the release of greenhouse gases, but the deforestation of tropical forests in Southeast Asia has a greater impact on the lives of people who depend on these forests for their livelihood

PART 2

The impact of human activities on the environment is felt around the world, and it can vary significantly depending on a variety of factors, including geography, culture, and economic development. In this chapter, we will examine the different types of human activities that have the greatest impact on the environment, as well as the regions and industries where the environmental impact is the most severe.

Agriculture: Agriculture is one of the oldest and most widespread human activities, and it has a profound impact on the environment. Agricultural practices such as deforestation, soil degradation, and the use of pesticides and fertilizers can cause a wide range of environmental problems, including loss of biodiversity, water pollution, and soil erosion. The impact of agriculture is particularly pronounced in developing countries, where subsistence farming is often the main source of food and income.

Transportation: Transportation is another key contributor to the human footprint on the environment. The burning of fossil fuels by cars, trucks, and other vehicles produces a significant amount of greenhouse gases and air pollution, which can have serious health and environmental consequences. The impact of transportation is felt most acutely in urban areas, where population density and heavy traffic contribute to high levels of air pollution and greenhouse gas emissions.

Energy production: The production and use of energy is another major contributor to the human footprint on the environment. Fossil fuels, such as coal, oil, and natural gas, are the main source of energy in many countries, and the extraction and burning of these fuels releases a large amount of greenhouse gases and air pollution. The environmental impact of energy production is particularly severe in developing countries, where energy infrastructure is often outdated and inefficient.

Urban development: Urbanization is a major contributor to the human footprint on the environment. As cities grow, they consume vast amounts of resources, including water, energy, and raw materials, and generate large amounts of waste. The impact of urbanization on the environment is most pronounced in developing countries, where rapid urban growth is often coupled with inadequate planning and infrastructure.

While the impact of human activities on the environment is felt around the world, it is particularly severe in developing countries, where limited resources and infrastructure can exacerbate the effects of environmental degradation. For example, in many developing countries, water pollution from agriculture and industry is a major problem, and it can have serious health consequences for the local population. In addition, deforestation in developing countries can cause a wide range of environmental problems, including soil erosion, loss of biodiversity, and climate change.

In conclusion, the human footprint on the environment is a complex and multifaceted issue, and it is shaped by a variety of factors, including geography, culture, and economic development. Understanding the different types of human activities that have the greatest impact on the environment, as well as the regions and industries where the impact is the most severe, is an important first step in addressing this critical issue.

Chapter 5: Mitigating Environmental Consequences

Introduction

The impact of human activities on the environment has been a growing concern for many years. Despite efforts to reduce the environmental footprint of human activities, the consequences of these actions continue to be felt all over the world. In this chapter, we will look at strategies for reducing the environmental impact of human activities, including conservation efforts, sustainable development, and renewable energy. We will also examine examples of communities and companies that have successfully reduced their environmental footprint.

Conservation Efforts

Conservation efforts refer to the preservation and protection of the environment through measures that are designed to reduce the impact of human activities. This can involve protecting natural habitats, reducing the use of natural resources, and promoting sustainable practices. For example, the creation of national parks and wildlife reserves helps to protect habitats and prevent the destruction of natural areas. In addition, programs such as the use of certified wood and paper products can reduce the impact of forestry practices on the environment.

Sustainable Development

Sustainable development is the development that meets the needs of the present without compromising the ability of future generations to meet their own needs. This involves the integration of economic, social, and environmental considerations into all aspects of development, including urban planning, transportation, and energy production. For example, the development of sustainable transportation systems, such as mass transit and bicycle-friendly infrastructure, can reduce the impact of transportation on the environment. In addition, the development of renewable energy sources, such as wind and solar power, can reduce the dependence on fossil fuels and their associated environmental impacts.

Renewable Energy

Renewable energy refers to energy sources that are replenished naturally and are not depleted over time. This includes sources such as wind, solar, hydro, and geothermal power. Renewable energy sources are considered to be more environmentally friendly than traditional sources such as coal, oil, and gas, as they do not produce harmful emissions or contribute to climate change. The use of renewable energy can also help to reduce the dependence on finite fossil fuels, which are becoming increasingly scarce and expensive.

Examples of Successful Environmental Mitigation Efforts

There are many examples of communities and companies that have successfully reduced their environmental footprint. For example, cities such as Amsterdam, Copenhagen, and Portland have developed sustainable transportation systems that encourage the use of mass transit and bicycles, reducing the impact of transportation on the environment. In addition, companies such as Patagonia, The Body Shop, and Tesla have implemented environmentally-friendly practices, such as using sustainable materials and reducing waste. These companies serve as examples of how businesses can have a positive impact on the environment while still maintaining profitability.

Conclusion

Reducing the environmental impact of human activities is a complex and ongoing challenge, but there are many strategies that can be implemented to achieve this goal. Conservation efforts, sustainable development, and renewable energy are all important strategies for mitigating the environmental consequences of human actions. By examining the successes of communities and companies that have successfully reduced their environmental footprint, we can learn from their examples and continue to find new and innovative ways to protect the environment for future generations.

Chapter 5: Looking to the Future

As we come to the end of this book, it is important to consider what the future holds for our planet and the environment. Our actions today will have a profound impact on future generations, and it is essential that we take a proactive approach to reducing our environmental footprint and ensuring a sustainable future. In this final chapter, we will explore what the future may hold for the environment, what we can do to ensure a sustainable future, and reflect on the key messages of the book.

The future of the environment: There is no doubt that the future of the environment is closely linked to the actions of human beings. Our continued reliance on fossil fuels, deforestation, and other activities that contribute to climate change and other environmental degradation are putting the planet at risk. It is essential that we take a proactive approach to reducing our environmental footprint and ensuring a sustainable future. There are many ways in which we can reduce our impact on the environment, including investing in renewable energy sources, reducing our use of single-use plastics, and conserving natural resources.

Ensuring a sustainable future: To ensure a sustainable future for the planet, we must focus on reducing our environmental footprint and adopting sustainable practices. This will require a collective effort from individuals, communities, and corporations alike. Governments must also play a key role in promoting sustainability and protecting the environment, by implementing policies and regulations that encourage sustainable practices. At the same time, businesses must take responsibility for their impact on the environment and adopt sustainable practices that reduce their footprint and promote environmental stewardship.

Reflecting on the key messages of the book: Throughout this book, we have discussed the environmental impact of human activities and the consequences that come with it. We have also explored the science behind these consequences and the history of human impact on the environment. In addition, we have looked at the human footprint in different regions and industries and explored strategies for mitigating environmental consequences. All of these topics have contributed to the overall message of this book, which is that we must take action to reduce our impact on the environment and ensure a sustainable future for the planet.

Recommendations for action: As we look to the future, it is important that we take action to reduce our impact on the environment. There are many ways in which individuals, communities, and corporations can make a positive impact, including investing in renewable energy sources, reducing our use of single-use plastics, and conserving natural resources. Governments must also play a key role in promoting sustainability and protecting the environment, by implementing policies and regulations that encourage sustainable practices.

In conclusion, the future of our planet and the environment is in our hands. We must take action to reduce our impact on the environment and ensure a sustainable future for the planet. By working together, we can create a better future for ourselves and future generations.

Part 2

Chapter 6: Looking to the Future

In the final chapter of "The Human Footprint: Navigating the Environmental Consequences of Our Actions," we will take a look at the future of human impact on the environment. The world is facing a number of environmental challenges, from climate change and loss of biodiversity to air and water pollution. These challenges are the result of human activities, and it is our responsibility to find solutions to mitigate their impact on the planet.

In this chapter, we will discuss the steps we can take to ensure a sustainable future for the planet. We will explore the concept of sustainable development, which aims to meet the needs of the present generation without compromising the ability of future generations to meet their own needs. This can include a variety of strategies, such as reducing our reliance on non-renewable resources, promoting renewable energy, and implementing conservation efforts.

We will also look at examples of communities and companies that have successfully reduced their environmental footprint. These examples can provide inspiration and guidance for others looking to make a positive impact on the environment. For example, we may discuss a company that has implemented a waste reduction program, a city that has implemented a public transportation system, or a community that has successfully implemented sustainable agriculture practices.

In addition to discussing practical solutions, we will also reflect on the key messages of the book and provide recommendations for action. We will emphasize the importance of taking individual and collective action to mitigate the environmental impact of our actions. We will encourage readers to consider their own impact on the environment and to take steps to reduce it.

In conclusion, this chapter will provide a hopeful outlook for the future, emphasizing the positive impact we can have on the environment through our collective efforts. We will remind readers that every small action can contribute to a more sustainable future for the planet and future generations. We will encourage readers to take responsibility for their own impact on the environment and to join us in working towards a more sustainable future.

www.ingramcontent.com/pod-product-compliance
Lightning Source LLC
Chambersburg PA
CBHW071148220526
45467CB00015B/2136